T·H·E HEAVY A·N·D LIGHT ANIMAL B·O·O·K

Kori Bustard
40 lb

Chihuahua
4 lb

Mouse Deer
7 lb

Goliath Bee
3 oz

Sea Otter
99 lb

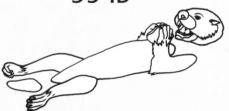

Anaconda
441 lb

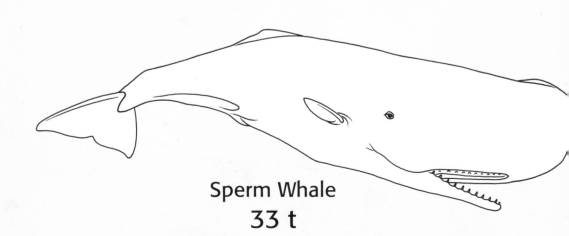

Sperm Whale
33 t

Kitti's Hog-Nosed Bat
.1 oz

African Pygmy Falcon
2 oz

Gecko
.03 oz

average weight of
an 8-year-old child
55 lb

Hippopotamus
3 t

Whale Shark
44 t

T·H·E
HEAVY
A·N·D
LIGHT
ANIMAL
B·O·O·K

DAVID TAYLOR
ILLUSTRATED BY
PETER MASSEY

RSVP

RAINTREE
STECK-VAUGHN
P U B L I S H E R S
The Steck-Vaughn Company

Austin, Texas

Contents

Words found in **bold** are explained in the glossary on page 31.

Published by Raintree Steck-Vaughn Publishers, an imprint of Steck-Vaughn Company

Illustration copyright © Peter Massey 1995
Text copyright © David Taylor 1995

Editor: Claire Edwards
Science Editor: Tracey Cohen
Electronic Production: Scott Melcer
Designer: Hayley Cove and Julie Klaus
Consultant: Fiona Collins

Library of Congress
Cataloging in Publication Data

Taylor, David, 1934–
    The heavy and light animal book / David Taylor; illustrated by Peter Massey.
       p.   cm. — (Animal opposites)
    Includes index.
    ISBN 0-8172-3953-7
    1. Animals—Juvenile literature. 2. Body weight—Juvenile literature. [1. Animals.] I. Massey, Peter, ill. II. Title. III. Series: Taylor, David, 1934– Animal opposites.
QL49.T2215   1996
591—dc20   95-11776
                      CIP   AC

Printed in Hong Kong
Bound in the United States
1 2 3 4 5 6 7 8 9 0 LB   99 98 97 96 95

Heavy and Light Animals

Which weighs more – a pound of lead or a pound of feathers? The weight is the same, of course! How much something weighs depends on what it is made of and how much there is of it. But what is weight?

The Force of Gravity

Weight is produced by a force called gravity. Gravity is the force of attraction between two objects in space, like the Earth and the moon. The pull of gravity from a huge object like the Earth gives weight to things on or near it. The Earth has a strong pull of gravity. Astronauts in a space rocket are weightless when they are far enough away from the Earth to escape the pull of gravity.

The Animal Kingdom

Animals come in an amazing range of weights. Tiny creatures, such as **amoebas**, each weigh less than a fraction of an ounce. Blue whales, the largest animals that have ever lived, can weigh up to 165 tons (150 mt). This is about 85 million times heavier than the Kitti's hog-nosed bat, one of the lightest living **mammals**. This tiny creature weighs about the same as three or four paper clips.

The heaviest human beings have weighed between 772 to 1,021 pounds (350 and 463 kg). The bigger an animal is, the heavier it will be. Also, bones are heavier than muscle, and muscle is heavier than fat. But some animals have special features that help cut down their weight. For example, birds have thin-walled, hollow bones that make them lighter.

Which Is Best?

A heavy animal has some advantages over a light animal. Few animals dare to attack an elephant. A heavy animal can carry larger stores of food within its body. The blue whale has tons of fatty **blubber** under its skin. This supplies it with energy during its long journeys across the world's oceans. But being light can also be useful. Light creatures need less food to survive. If they are small they can escape from enemies by hiding in cracks and other tiny spaces.

It is true that the fastest animals in the world are not the biggest **species**. But some large animals, such as horses and killer whales, have bodies with plenty of muscle and can move at high speeds.

Very small, very light animals may be supported by air. Spiders use the silk they spin as a simple kite. This helps them sail through the sky, carried by currents of air, for hundreds of miles.

This book looks at creatures that are lightweights and others that are super-heavyweights. All of them have an equally important place on Earth.

How Do We Weigh Things?

People have weighed things, using scales or balances, for thousands of years. In the past, if someone wanted to weigh something, they would compare its weight with that of a special piece of metal or stone. This then became a standard measurement of weight. The earliest units of weight were called *beqas*. They were short **cylinders** with rounded ends, made of stone. *Beqas* were used in Egypt around 8000 to 7000 BC.

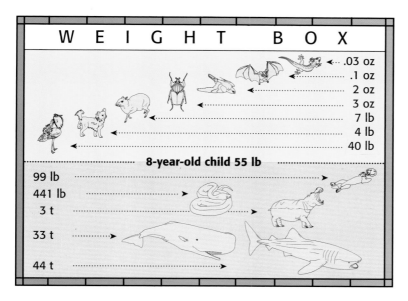

WEIGHT BOX

These weight boxes give an idea of how much the animals in this book weigh compared to each other. The average weight of an eight-year-old child is also included. The abbreviations in the box are g (gram), kg (kilogram), and metric ton (mt).

In the United States, the most common units for measuring weight are ounces, pounds, and tons. There are 16 ounces in a pound, and 2,000 pounds in a ton. But we also use metric units—metric tons, kilograms, and grams, and, even smaller, milligrams.

Where in the World?

You can see the places where the animals talked about here live. Turn to pages 30 and 31.

The Whale Shark

The super-heavyweight champion of the world's sharks is not the great white shark of the *Jaws* movies. It is the whale shark. This harmless monster of a fish that can grow to about 60 feet (18 m) in length and weigh as much as 44 tons (40 mt). It has gray-brown skin that is decorated with a pattern of pale lines and yellow spots. From tip to tip, its tail is taller than a bull, or male, elephant.

Although it is huge, the whale shark is a gentle creature. It never attacks humans, although once in a while it might upset a fishing boat by accident, as it rises to the surface.

Like the blue whale, the other giant of the oceans, the whale shark feeds on millions of tiny animals called plankton. The shark gulps down a huge mouthful of water and plankton, clamps its jaws shut, and pumps the water out through its **gills**. In the gills there are rows of brushes, called gill rakers. These filter out the food, so that the whale can swallow it.

Whale shark

Prehistoric Shark

Fossil teeth have been found on the seafloor
from sharks that lived millions of years ago.
One creature was like a giant great
white shark and could grow up
to 98 feet (30 m) long.

A Shy Creature

The whale shark was first
discovered in the early 1800s.
People have seen it less than 200
times since then. It lives in warm
waters and has mostly been seen in
the Straits of Florida, the Caribbean,
and around the Philippines. A shark
may sometimes be seen as far
north as New York.

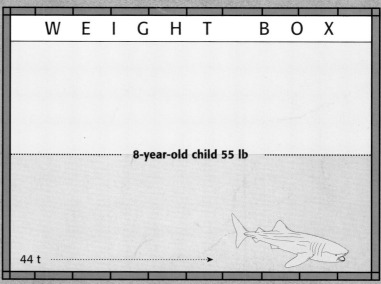

WEIGHT BOX

.... **8-year-old child 55 lb** ....

44 t ....➤

The Sperm Whale

The sperm whale is one of the heaviest mammals that has ever lived on Earth. It is the biggest of the toothed whales. This whale is very intelligent and has a brain that weighs about 21 pounds (9.5 kg). Male sperm whales can weigh 44 tons (40 mt), which is more than six Indian elephants.

The sperm whale has a certain kind of liquid fat in its head. It uses this liquid fat to help it dive down, or rise to the surface of the ocean. The whale cools the fat using air tubes linked to its **blowhole**. The whale fills these tubes with water, which cools the fat and turns it into hardened wax. The wax is heavier than the liquid fat and helps the whale dive. When the whale wants to swim to the surface, extra blood flows around its head and warms the wax. This turns it back to its lighter liquid form. Like all whales, dolphins, and porpoises, sperm whales have lungs and breathe air, but they can make deep, hour-long dives. After these, they need to rest and stay on the surface for 30 minutes or more. They can be found in oceans all around the world.

Weighing a Whale

This method of weighing a whale can only be done for whales up to the size of the killer whale. Scientists put the whale in a stretcher. Then the stretcher is attached to a scale. The scale is hung from a big crane, which lifts the animal so that it can be weighed. Using this method the animal is not harmed, and scientists can learn the weight of the whale.

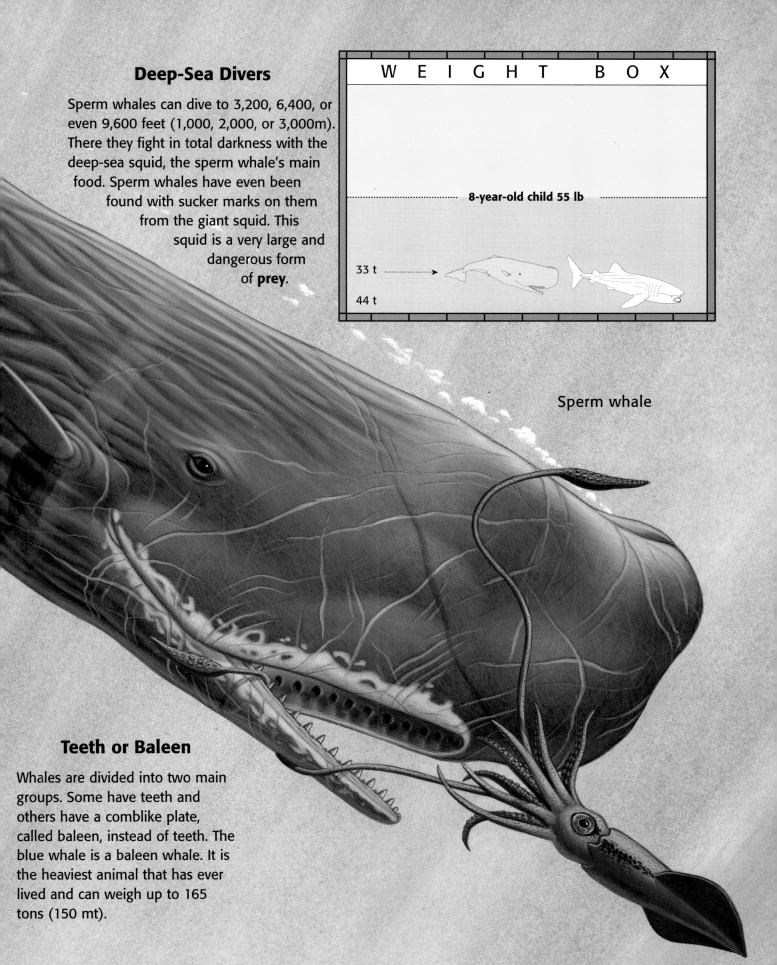

Deep-Sea Divers

Sperm whales can dive to 3,200, 6,400, or even 9,600 feet (1,000, 2,000, or 3,000m). There they fight in total darkness with the deep-sea squid, the sperm whale's main food. Sperm whales have even been found with sucker marks on them from the giant squid. This squid is a very large and dangerous form of **prey**.

W E I G H T   B O X

8-year-old child 55 lb

33 t

44 t

Sperm whale

Teeth or Baleen

Whales are divided into two main groups. Some have teeth and others have a comblike plate, called baleen, instead of teeth. The blue whale is a baleen whale. It is the heaviest animal that has ever lived and can weigh up to 165 tons (150 mt).

The Hippopotamus

Hippopotamuses spend most of their lives in the water. They can weigh over 3.5 tons (3.2 mt). When they are covered with water in an African river, the water supports their bodies. Then they weigh almost nothing. Underwater they walk elegantly along the muddy bottom. But it's wise not to go too close, especially if there are young nearby on the bank. Female hippos raise their young on the mud banks. Although it looks peaceful enough, a hippo mother can be very dangerous. If disturbed or frightened, it will suddenly charge, and chomp with its mighty jaws at anything that might be a threat. Hippos have been known to turn over small boats and bite people.

River Horse

The name *hippopotamus* comes from the ancient Greek word meaning "river horse." With its long head, broad mouth, and flicking ears, the hippo does look just like a fat horse!

Hippopotamus

Red Sweat

People once believed that hippopotamuses sweated blood. In fact, their sweat is a thick, oily, pinkish liquid. It keeps their skin becoming too dry when they are out of the water.

Where to Find Hippos

There are two kinds of hippopotamuses. The common hippopotamus was once found throughout Africa in rivers and lakes surrounded by grasslands. People killed it in very great numbers, so it is now rare in large parts of Africa. The smaller pygmy hippopotamus lives in forests and is found only in a few parts of West Africa. This pygmy hippo is in danger of extinction. Both kinds of hippos are **herbivores**. They eat grass and other plants.

W E I G H T   B O X

8-year-old child 55 lb

3 t

33 t

44 t

The Anaconda

All snakes are **carnivores**. Some kill their prey by injecting poison into the prey through hollow teeth called fangs. Others coil around their prey and squeeze their prey so tightly that it cannot breathe. The heaviest snakes in the world are all of the second kind. The reticulated python of tropical Asia, the African python, and the boa constrictor of South and Central America are all long, very heavy snakes. But the heaviest of all these snakes is the anaconda, which lives in swamps and slow-running rivers in South America.

A Big Appetite

Early Spanish settlers in South America believed that the anaconda swallowed human beings and cattle. This is unlikely. But it does kill and swallow whole big creatures, such as pigs, small deer, and birds.

All Muscle

The anaconda has huge, powerful muscles that run along the length of its body. It uses these muscles to squeeze and kill its prey.

The anaconda can grow to 31 feet (9.6 m) in length and weigh over 441 pounds (200 kg). Its skin is olive-brown with two rows of large black oval spots running down its back and smaller white spots along its sides. The anaconda eats mainly at night and spends most of its time in the water. Only a small part of its head shows above the surface as it waits to catch its prey.

Anaconda

W E I G H T	B O X

···· **8-year-old child 55 lb** ····

441 lb
3 t

33 t

44 t

The Sea Otter

The sea otter is the lightest sea mammal. It is lighter than the smallest dolphin or porpoise. Males, which are about one-third heavier than females, weigh between 49 and 99 pounds (22 and 45 kg). That is about 3,000 times lighter than the heaviest sea mammal, the blue whale. Sea otters live along parts of the California coast, in the Gulf of Alaska, and around the Kuril and Aleutian islands.

A sea otter spends most of its life in the ocean. It eats slow-moving fish and animals. The otter often dives underwater to search for food.

Keeping Warm

Whales, dolphins, and porpoises have a layer of fat beneath their skins. This is called blubber, and it helps to keep out the cold. Sea otters do not have blubber. Instead they have thick furry coats that trap bubbles of air. Because air does not carry heat very well, these bubbles reduce heat loss and keep the otters warm.

A sea otter's paws are good for grasping. When it has caught something, it lies on its back on the ocean's surface. Then the otter cracks the shell open by hitting it against a stone that rests on its chest. The only other mammals that use tools in this way are chimpanzees—and humans.

Claws That Clean

Sea otters are always grooming themselves to keep their coats fluffed up with air. Their front paws are good for grooming. Sea otters have claws that can be pulled in like a cat's claws.

Sea otter

W E I G H T	B O X
8-year-old child 55 lb	
99 lb	
441 lb	
3 t	
33 t	
44 t	

The Kori Bustard

The heaviest living bird is the ostrich, but the ostrich cannot fly. The heaviest flying bird is the Kori bustard. There are 21 species of bustards around the world. Some of these are quite small. Others are the heaviest of all flying animals. The male Kori bustard that lives in Africa weighs up to 40 pounds (18 kg)— as much as a terrier dog. If it weighed much more than this, it would not be able to fly. When the Kori bustard does fly, it needs a long run to get itself up into the air. It can fly only about 328 feet (100 m) before it has to land.

Other heavy bustards are the great bustard of Europe, Africa, and Central Asia; the Australian bustard; and the rare and endangered great Indian bustard.

Well-Hidden

All bustards have long necks and legs. They have brown or reddish-brown feathers with black markings. This helps **camouflage**, or hide, them in the deserts and grasslands where they live.

Lying Low

The name bustard means "bird that walks" in Latin. Bustards are nervous birds that live on the ground. They are strong, slow walkers that seek cover at the first sign of danger. When attacked by an enemy, they can fly very quickly.

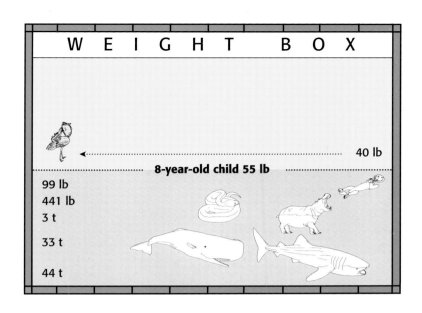

W E I G H T   B O X

40 lb

8-year-old child 55 lb

99 lb
441 lb
3 t
33 t
44 t

A Mixed Diet

Kori bustards eat seeds, berries, insects, small reptiles, **rodents**, and the eggs of other birds that nest on the ground.

Kori bustard

Hunters

Chihuahuas just look weak and fragile. Although they no longer live in the wild, they are good hunters. They also adjust well to any climate.

A Dog-Rabbit

There are two kinds of modern Chihuahuas. One has smooth coat with short, glossy fur, and one with long coat and longer hair. People once thought the Chihuahua was a cross between a small dog and a rabbit, or a kind of rodent called a prairie dog. But such cross-breeding would not be possible.

Chihuahua

The Chihuahua

The Chihuahua is the smallest and lightest breed of tame dogs. It is one of a group of very small dogs called toy dogs that are now bred as pets. The Chihuahua weighs between 1 and 6 pounds (.5 and 3 kg). This is up to 33 times less than the heaviest breed of dogs, the Saint Bernard. The Chihuahua came to the United States from Mexico in 1898. But its history before then is a mystery. This tiny dog may be an ancestor of the dogs honored by the **Toltec** people. The **Aztecs** may have raised them for meat and for their soft coats that were made into cloth. Perhaps they were introduced into South America from Spain in the 16th century, or from China in the 19th century. The Chihuahua probably comes from both ancient and modern breeds of dogs.

WEIGHT BOX

4 lb
40 lb

8-year-old child 55 lb

99 lb
441 lb
3 t

33 t

44 t

The Mouse Deer

The lesser mouse deer looks like a deer but is about the size of a rabbit. It weighs only 5 to 10 pounds (2 to 5 kg). But it isn't really a deer at all. It lives in the rain forests of Southeast Asia. The mouse deer belongs to an ancient family of animals called the chevrotains. These animals have not changed much over the past 30 million years. They and are probably related to pigs and camels.

Mouse deer are ruminants. This means that they belong to a group of animals that store food in a stomach made up of four compartments. They digest or break their food down slowly with the help of millions of friendly **microbes**. They have no horns or antlers. Males have two fangs that stick out at the sides of their mouths. These fangs never stop growing, but they are being worn down all the time.

At Risk

The mouse deer is a shy animal. It sleeps during the day. At night it looks for fallen fruit and the leaves of low-growing plants to eat. It has many enemies and is hunted by eagles, wild cats, large snakes, and crocodiles. People hunt the mouse deer and are also slowly destroying its **habitat**.

Real Deer

The smallest deer in the world is the pudu of northern South America. It weighs between 13 and 30 pounds (5.8 and 13.4 kg), making it about twice the size of the lesser mouse deer.

Keeping In Touch

Mouse deer usually keep to themselves. When they do meet, they call to each other. They also know each other by their smells.

Lesser mouse deer

WEIGHT BOX

7 lb
4 lb
40 lb

8-year-old child 55 lb

99 lb
441 lb
3 t

33 t

44 t

The Goliath Beetle

The Goliath beetle, named after the giant Goliath, is one of the heaviest insects in the world. This amazing giant beetle has black and white stripes on the front half of its body. It is more than 80 times bigger than a ladybug. The Goliath beetle lives in forest areas of Africa near the Equator. Male Goliath beetles can weigh up to 3.5 ounces (100 g), which is about as heavy as a lemon. Imagine a creature as big as a hamster and as heavy as a lemon buzzing into your face! But the Goliath beetle only flies if it has to, and then very slowly. It doesn't sting or bite and eats only soft fruit and sap from trees.

Goliath beetles hang from trees in the same way monkeys do. They wrap their strong legs around a small branch and hang on tightly. If you try to pull one away, it will scratch you with its hard claws.

Rival Heavyweights

Other heavy insects include some stick insects, the elephant beetle, and longhorn beetles. Both elephant and longhorn beetles are found in South America. The people who live along the Amazon River eat the grubs, or larvae, of Longhorn beetles. They are supposed to be delicious!

Bright Lights

The giant titan longhorn beetle of Brazil is even heavier than the Goliath beetle. This rare beetle can weigh up to 5.25 ounces (150 g) and is attracted to bright light. It has been seen crawling on the ground beneath street lamps in small towns along the Amazon River.

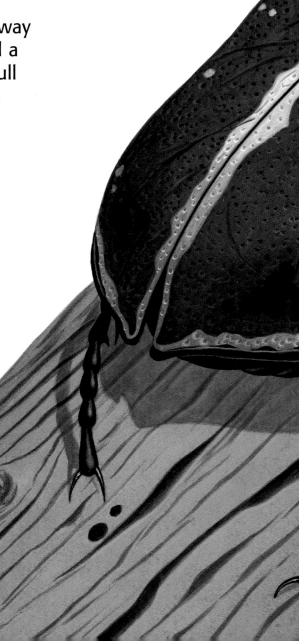

Goliath beetle

W E I G H T   B O X

3 oz
7 lb
4 lb
40 lb

8-year-old child 55 lb

99 lb
441 lb
3 t

33 t

44 t

The African Pygmy Falcon

The greatest hunters of the bird world are the birds of prey. This group includes the largest of all flying birds: condors, vultures, eagles, and elegant falcons. These birds are built for finding their prey, holding, killing, and tearing it apart. They have good vision, strong feet with curved claws, and powerful beaks.

This group of hunters also includes the smallest hunting birds, the Philippine falconet and the African pygmy falcon. The word *pygmy* means "small." The African pygmy falcon weighs only 1.4 to 2.1 ounces (40 to 60 g). That is no more than a sparrow.

African pygmy falcon

Nest Stealer

In South Africa the pygmy falcon breeds in the nests of the little seed-eating birds called social weavers. In East Africa, it lives among the Buffalo weavers, in their strong nests made of thorny twigs.

Heavy Hunter

The condor of the South American Andes Mountains is the heaviest bird of prey. It weighs between 20 and 24 pounds (9 and 11 kg). Its wings measure 9.8 feet (3m) from tip to tip. If you imagine how long a small car is, you will have an idea of its wingspan.

The African pygmy falcon looks more like a finch than a bird of prey. But, like the peregrine falcon, it is a fierce, clever hunter. It lives mainly on large insects and small birds.

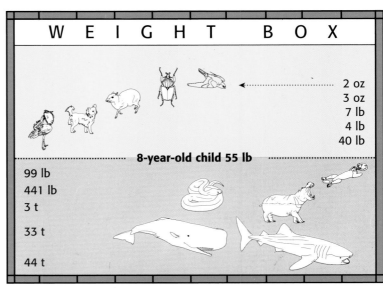

W E I G H T   B O X

2 oz
3 oz
7 lb
4 lb
40 lb

8-year-old child 55 lb

99 lb
441 lb
3 t
33 t
44 t

The Kitti's Hog-Nosed Bat

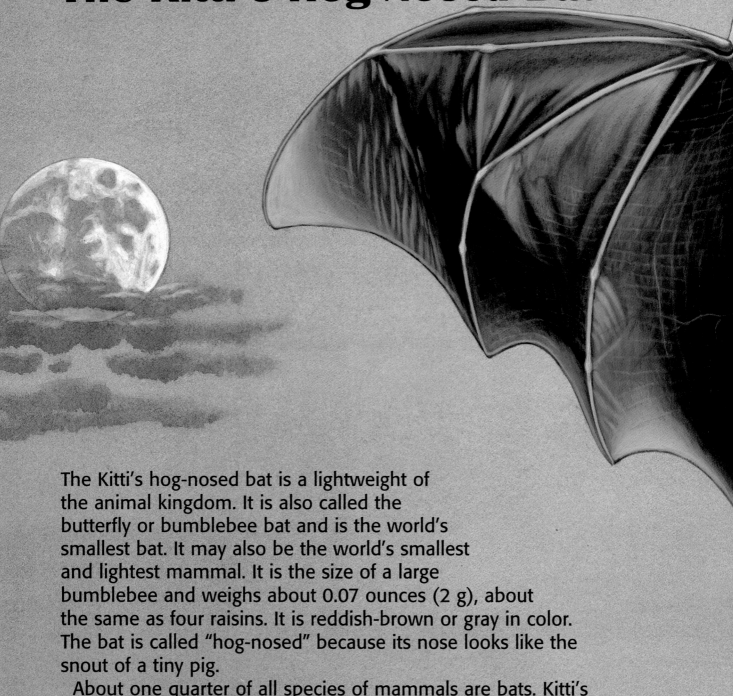

The Kitti's hog-nosed bat is a lightweight of the animal kingdom. It is also called the butterfly or bumblebee bat and is the world's smallest bat. It may also be the world's smallest and lightest mammal. It is the size of a large bumblebee and weighs about 0.07 ounces (2 g), about the same as four raisins. It is reddish-brown or gray in color. The bat is called "hog-nosed" because its nose looks like the snout of a tiny pig.

About one quarter of all species of mammals are bats. Kitti's hog-nosed bats live in colonies or large groups. The colonies are usually small, and the bats roost away from each other, deep inside limestone caves. Most bats eat insects, but some eat fruit, pollen, nectar, small mammals, frogs, lizards, or fish. A few, like the vampire bat, make a tiny cut and lap the blood of much bigger animals. The Kitti's hog-nosed bat is one of the 650 species of bats that eats insects.

Hunting at Dusk

Kitti's hog-nosed bats leave their cave at dusk to go hunting. They send out high-pitched sounds that bounce back from objects as echoes. Their large ears pick up these echoes that help them catch their prey.

Recent Discovery

The Kitti's hog-nosed bat was discovered in 1973, living in the forests of Thailand. Because people have disturbed its habitat it is now an endangered species. Endangered species are those that are in danger of dying out. There are only about 200 of these tiny bats left in the world.

Kitti's hog-nosed bat

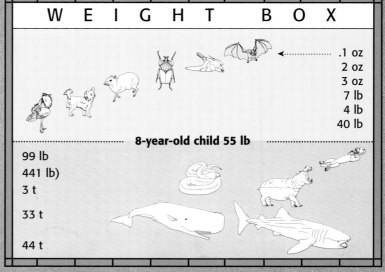

W E I G H T B O X

.1 oz
2 oz
3 oz
7 lb
4 lb
40 lb

8-year-old child 55 lb

99 lb
441 lb)
3 t

33 t

44 t

The Gecko

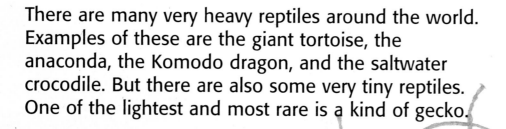

There are many very heavy reptiles around the world. Examples of these are the giant tortoise, the anaconda, the Komodo dragon, and the saltwater crocodile. But there are also some very tiny reptiles. One of the lightest and most rare is a kind of gecko.

It does not have an English name, just a long Latin one: *Sphaerodactylus parthenopion*. It is just less than 2 inches (4 cm) long and weighs no more than a fraction of an ounce (1g), about the same as a couple of shelled peanuts. It was discovered during the 1960s in the Virgin Islands in the Caribbean. Little is known about this tiny lizard.

Geckos have soft, scaly skin, short fat bodies, and large heads. They live in warm climates and are usually gray, brown, or dirty white. Geckos change color to blend in with their surroundings.

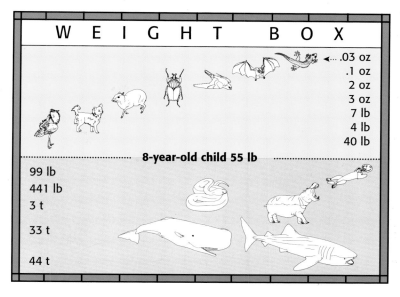

W	E	I	G	H	T	B O X

.03 oz
.1 oz
2 oz
3 oz
7 lb
4 lb
40 lb

8-year-old child 55 lb

99 lb
441 lb
3 t
33 t
44 t

Velvet Feet

The bottoms of geckos' feet are like velvet. They are covered with tiny hook cells that are like bristles on a brush. These hooks help the gecko grab onto surfaces and let it run up walls and even across ceilings.

Gecko

A Bad Name

In some countries people think geckos have a poison in their bite and poison on their skin. In fact they are harmless creatures. They are good to have around the house because they eat insects.

Snakes' Eyes

Geckos don't have eyelids. Like snakes, they have a transparent, protective covering over their eyes.

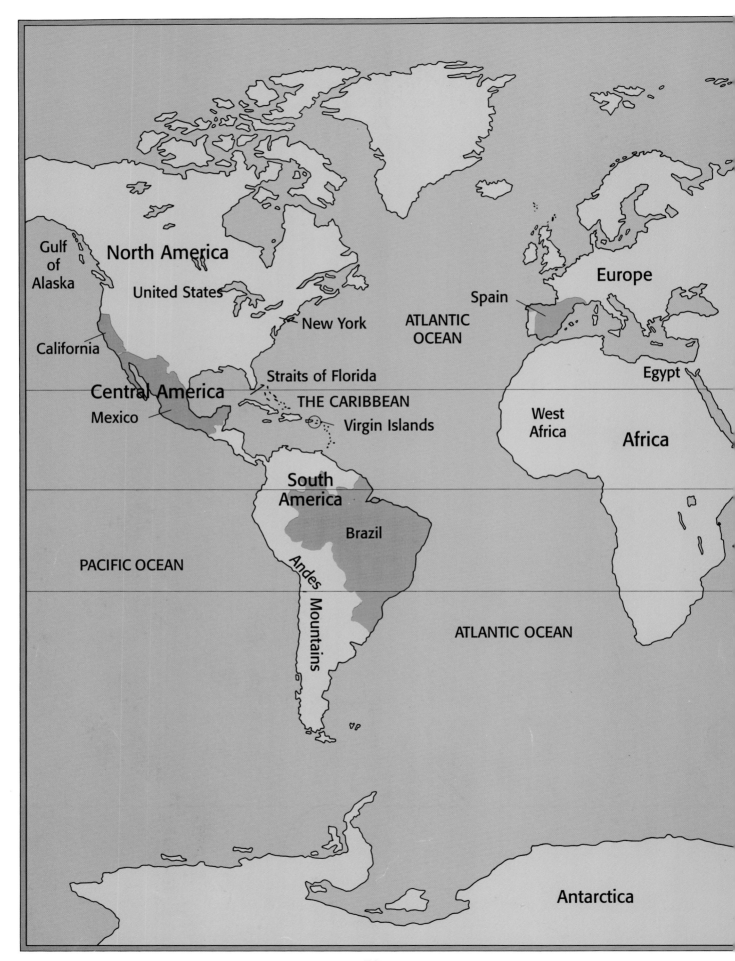

Gulf of Alaska

North America

United States

California

New York

ATLANTIC OCEAN

Spain

Europe

Central America

Straits of Florida

THE CARIBBEAN

Mexico

Virgin Islands

Egypt

West Africa

Africa

South America

Brazil

PACIFIC OCEAN

Andes Mountains

ATLANTIC OCEAN

Antarctica

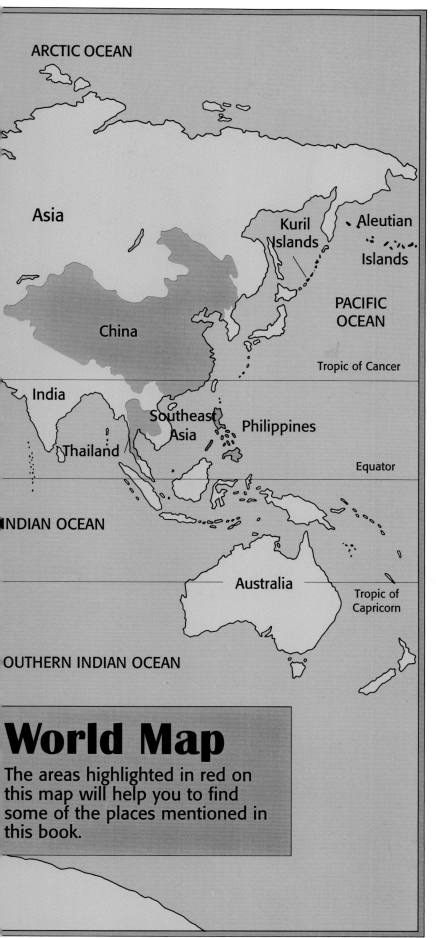

World Map

The areas highlighted in red on this map will help you to find some of the places mentioned in this book.

Glossary

amoeba A microscopic animal creature.

Aztecs and Toltecs Two ancient peoples from the part of the Americas now known as Mexico.

blubber The layer of fat beneath the skin of many animals that live in the ocean.

blowhole The opening through which a whale, dolphin, or porpoise breathes air.

camouflage The colors or pattern on an animal that make it difficult to see against its background.

carnivore An animal that eats other animals or flesh as food.

cylinder A solid or hollow tube that has round ends that are the same size.

fossil The remains of a dead animal or plant that has changed into stone over thousands of years.

gills Thin sheets of cells that take in oxygen from water. Fish use gills instead of lungs to breathe.

habitat The place in which a plant or animal usually lives.

herbivore A plant-eating animal, one that eats only plants.

microbe A tiny form of life. One example of a microbe or microorganism is bacteria. Friendly bacteria help us to digest food.

prey Animals or other living things that are killed by other animals for food.

species Groups of animals and plants that are similar to each other and have similar features.

Toltecs (*see* **Aztecs**).

Index